The
Big
Water
Puddle
(A Look at the Water Cycle)

Teacher Resource Companion
Christen McCray-Moore

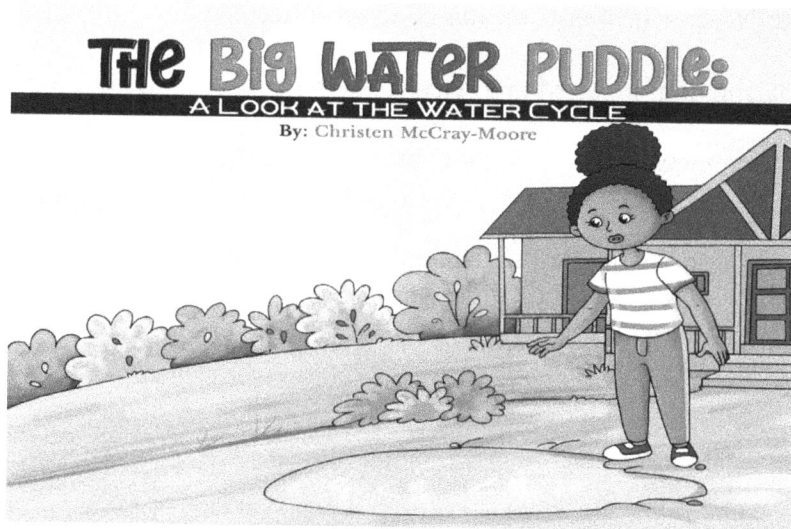

THE BIG WATER PUDDLE:
A LOOK AT THE WATER CYCLE
By: Christen McCray-Moore

Copyright

The Science Connection

Teacher Resource Companion: The Big Water Puddle: A Look at the Water Cycle © 2020

Christen McCray-Moore

Printed in the United States of America

ISBN: 978-1-7336134-3-9

Publisher: Armor of Hope Writing & Publishing Services, LLC.

Cover Design: Denise M. Walker

Table of Contents

The Big Water Puddle

Purpose

The purpose of this resource companion is to give teachers and/or parents, strategies and activities to reinforce the science content that coincides with my children's book, "The Big Water Puddle (A Look at the Water Cycle)". This companion includes the following: pre-reading, during reading, and after reading strategies and additional cross curriculum activities.

The Big Water Puddle

Fl. Science Standards

SC.5.E.7.1- Create a model to explain the parts of the water cycle. Water can be a gas, a liquid, or a solid and can go back and forth from one state to another.

SC.5.E.7.2- Recognize that the ocean is an integral part of the water cycle and is connected to all of Earth's water reservoirs via evaporation and precipitation processes.

FL. ELA Standards

Strand: Writing Standard

Cluster 1: Text Types and Purposes

LAFS.5.W.1.2 Write informative/explanatory texts to examine a topic and convey ideas and information clearly. e. Provide a concluding statement or section related to the information or explanation presented.

Strand: Language Standards

Cluster 3: Vocabulary Acquisition and Use

LAFS.5.L.3.6 Acquire and use accurately general academic and domain-specific words and phrases as found in grade level appropriate texts, including those that signal contrast, addition, and other logical relationships (e.g., however, although, nevertheless, similarly, moreover, in addition).

GA. Science Standards

S4E3 Obtain, evaluate, and communicate information to demonstrate the water cycle. a. Plan and carry out investigations to observe the flow of energy in water as it changes states from solid (ice) to liquid (water) to gas (water vapor) and changes from gas to liquid to solid. b. Develop models to illustrate multiple pathways water may take during the water cycle (evaporation, condensation, and precipitation).

GA. ELA Standards

Reading Literary

Key Ideas & Details-**ELAGSE4RL3**- Describe in depth a character, setting, or event in a story or drama, drawing on specific details in the text (e.g., a character's thoughts, words, or actions).

Language

Vocabulary Acquisition and Use-ELAGSE4L4 - Determine or clarify the meaning of unknown and multiple-meaning words and phrases based on grade 4 reading and content, choosing flexibly from a range of strategies. a. Use context (e.g., definitions, examples, or restatements in text) as a clue to the meaning of a word or phrase. b. Use common, grade-appropriate Greek and Latin affixes and roots as clues to the meaning of a word (e.g., telegraph, photograph, autograph).

National Standards

Earth and Space Science

Content Standard D

As a result of their activities in grades 5-8, all students should develop an understanding of

*Structure of the earth system

*Earth's history

*Earth in the solar system

The Big Water Puddle

Water Cycle Lesson Plan (Multiple days)

Objective: The students will (TSW) explain and create a model of the water cycle. TSW recognizes the ocean is essential to the water cycle.

Key Vocabulary: precipitation, condensation, evaporation, run-off, meteorologist, infiltration, transpiration

Fl. Science Standards: SC.5.E.7.1, SC.5.E.7.2

ESE/ESOL Strategies: extra help, oral strategies, visual aids, peer tutoring, cooperative learning and the use of graphic organizers

Materials: KWL charts, pencils, chart paper, zip lock bag or dome shaped bowl with a clear lid, colored ice cube, food coloring

Preparation: Put water into an ice cube tray, add food coloring and place into the freezer until frozen (to be used for the inquiry activity).

Connection Activity

Explain the various parts of the KWL chart. Have students create individual KWL charts and fill in the **"K"** with what they know about the water cycle. The teacher will have students share what they know from their individual KWL charts. The teacher will combine their thoughts/comments to create a class KWL chart on the water cycle (the chart should be written on chart paper or another large piece of paper).

Investigative Activities

1. Students will fill in the **"W"** with what they want to know about the water cycle. The teacher will have students share what they know from their individual KWL charts. The teacher will combine their thoughts/comments onto the previously created class KWL chart.

2. Students will participate in an inquiry activity. A student volunteer will draw clouds at the top of a zip lock bag and place a colored ice cube into it or dome shaped bowl with a clear lid. The student will place the zip lock bag or container under a bright lamp for 30 minutes. Meanwhile, students will complete the water inquiry sheet. Students will predict what will happen to the ice cube and infer how the model that was just created is similar to the way water moves through the water cycle.

The Big Water Puddle

Water Cycle Lesson Plan

Enlightenment Activities

1. The teacher and students will do a preview of the book, "The Big Water Puddle (A Look at the Water Cycle)". Ask probing questions: What do you think will occur in the story? Do you think the main character will have a conflict in the story? If so, what will it be? Discuss vocabulary words. The teacher will read the book, stopping at various points to ask comprehension questions (see pg. 8).

Extension Activities

1. Refer back to the ice cube in the zip lock bag or container and ask how their thoughts may or may not have changed and clarify any misconceptions.

2. Have students participate in various projects/assignments (see pgs. 10-19).

Group Project-Research/create water cycle on presentation board

Diorama-Create a 3D model of the water cycle

"Role Playing"-Act out the steps of the water cycle

Extension-Add a character to the story

Writing Prompt-Explain how you would teach the water cycle to someone

Assessment Activities

1. Ask comprehension questions (see pg. 9)

2. Students will complete the **"L"** part of the KWL that they started prior to reading the book. Students will complete their individual KWL chart. Teacher will have students share what they learned from their individual KWL chart and combine their thoughts/comments to complete the class KWL chart.

3. Choose various activities to test student's knowledge of the water cycle.

Story Map- Outline essential story elements

Individual Sequencing Map- Students will design their own sequence map of the water cycle.

Water Inquiry

Directions: Draw a picture of what the water looks like at the beginning and all throughout the activity. Fill in the sentences explaining your observations. Infer how water moves through the water cycle.

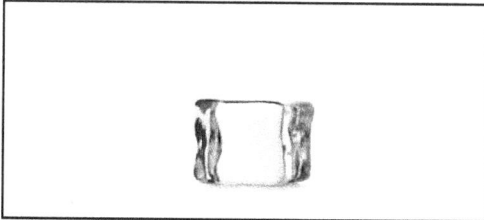

1. The water is in the form of an ice cube, so it is in _____ form.

2. The lamp is causing the ice cube to _____ up and _____, changing from a solid to a liquid.

3. Some of the water is collecting into the bottom of the bag representing "_____".

4. Some of the water is _____, changing from liquid to gas/water vapor due to the light/heat source.

5. Tiny water _____ are forming in the bag simulating _____. Condensation turns water vapor into a liquid.

6. When the water droplets become heavy, they fall back down as _____ and the cycle repeats.

Water Inquiry - Answers

Directions: Draw a picture of what the water looks like at the beginning and all throughout the activity. Fill in the sentences explaining your observations. Infer how water moves through the water cycle.

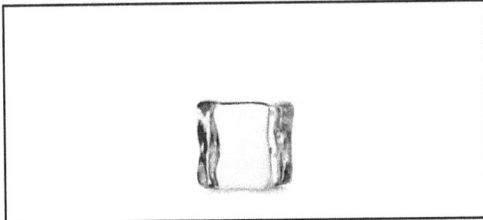

1. The water is in the form of an ice cube, so it is in _solid_ form.

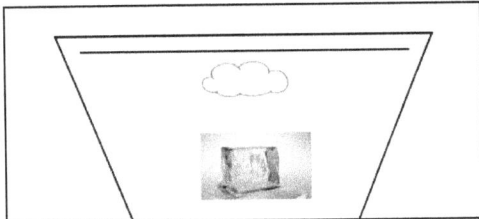

2. The lamp is causing the ice cube to _heat_ up and _melt_, changing from a solid to a liquid.

3. Some of the water is collecting into the bottom of the bag representing "_run off_".

4. Some of the water is _evaporating_, changing from liquid to gas/water vapor due to the light/heat source.

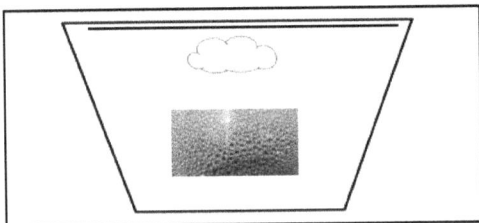

5. Tiny water _droplets_ are forming in the bag simulating _condensation_.
Condensation turns water vapor into a liquid.

6. When the water droplets become heavy, they fall back down as _precipitation_ and the cycle repeats.

The Big Water Puddle
Pre-Reading Strategies
Vocabulary and KWL

Vocabulary

Precipitation	Evaporation	Condensation
Run-Off	Infiltration	Transpiration
	Meteorologist	

1. Discuss vocabulary (showing vocabulary cards) prior to reading the book The Big Water Puddle (A Look at the Water Cycle). Ask students if they are familiar with any of the words or if they know what they mean. Give students background info on each unfamiliar word.

"KW" L Chart

1. Explain the various parts of the KWL chart. Have students create individual KWL charts and fill in the K-what they know and the W-want to know about the water cycle.

2. The teacher will have students share what they know and want to know from their individual KWL charts. The teacher will combine their thoughts/comments to create a class KWL chart on the water cycle (the chart should be written on chart paper or another large piece of paper).

The Big Water Puddle

Vocabulary Cards

Precipitation	Evaporation
Condensation	Run Off
Infiltration	Transpiration
Meteorologist	

Name _____

Date _____

K now

W ant to know

L earned

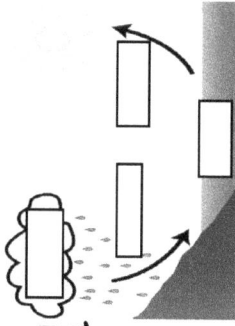

The Big Water Puddle

During Reading Strategies

Comprehension Questions

Comprehension Questions

1. The teacher will ask comprehension questions throughout the story.

* Who is the main character in the story?

* As soon as Dee Dee and her family moved into their new home, she set out to go on an adventure. What did she want to explore?

* Soon Dee Dee becomes disappointed, why?

*What is evaporation?

* What occurs during condensation?

*What is precipitation, and what form of precipitation is described in the story?

* Compare infiltration and run-off.

The Big Water Puddle

After Reading Strategies
Comprehension Questions & KW "L"

Comprehension Questions

1. The teacher will ask comprehension questions about the story.

*At the end of the story, Dee Dee explained what she learned about the water cycle. Explain in your own words how the water cycle works?

*Why is the water cycle important?

* Are there any parts of the water cycle we can leave out? Why not?

* What part does the ocean play in the water cycle?

* What is the term for when water is evaporated from the leaves of plants and trees?

KW "L" Chart

1. Students will complete the "L" part of the KWL that they started prior to reading the book. Students will complete their individual KWL chart.

2. The teacher will have students share what they learned from their individual KWL chart. The teacher will combine their thoughts/comments to complete the class KWL chart that was previously started on the water cycle.

The Big Water Puddle (A Look at the Water Cycle)

Story Map

Setting

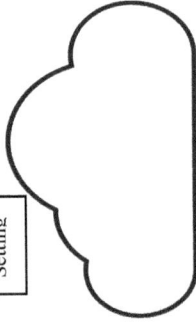

Solution

Plot

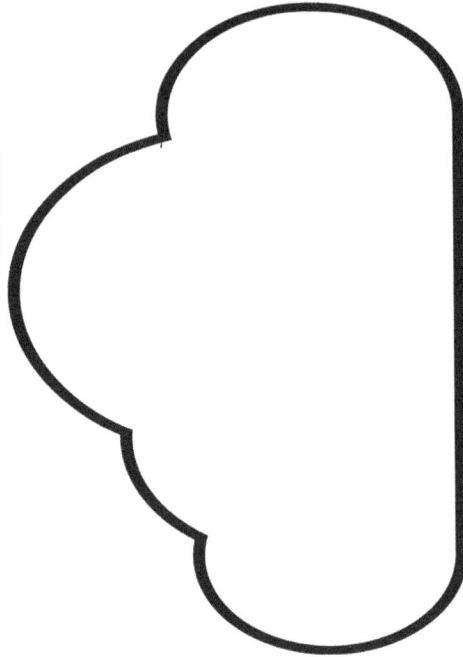

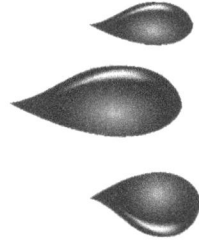

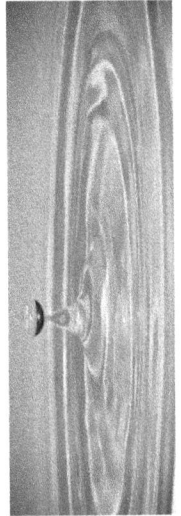

Problem

Characters

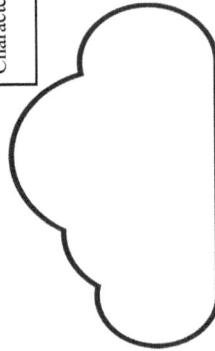

The Big Water Puddle

Closure Activities
Role Playing

Role Playing

After students read and discuss the story, have them 'act out" the steps of the water cycle. Give various students a part of the water cycle (picture cutouts of precipitation-rain, condensation-cloud, etc.), and they will pretend to "act out" that part of the water cycle. Next, place it on the dry erase board in the proper sequence. As students are acting out the role on their card, students will give hints by telling what they are doing. For example, if the card says evaporation, a student could be seen first squatting on the floor and standing up (with arms stretched to the ceiling) After the activity, allow students to discuss what they learned through a class discussion.

Note-Picture cutouts could be copied, put on cardstock, laminated with magnets or another adhesive placed on the back.

condensation

sun

ocean

precipitation

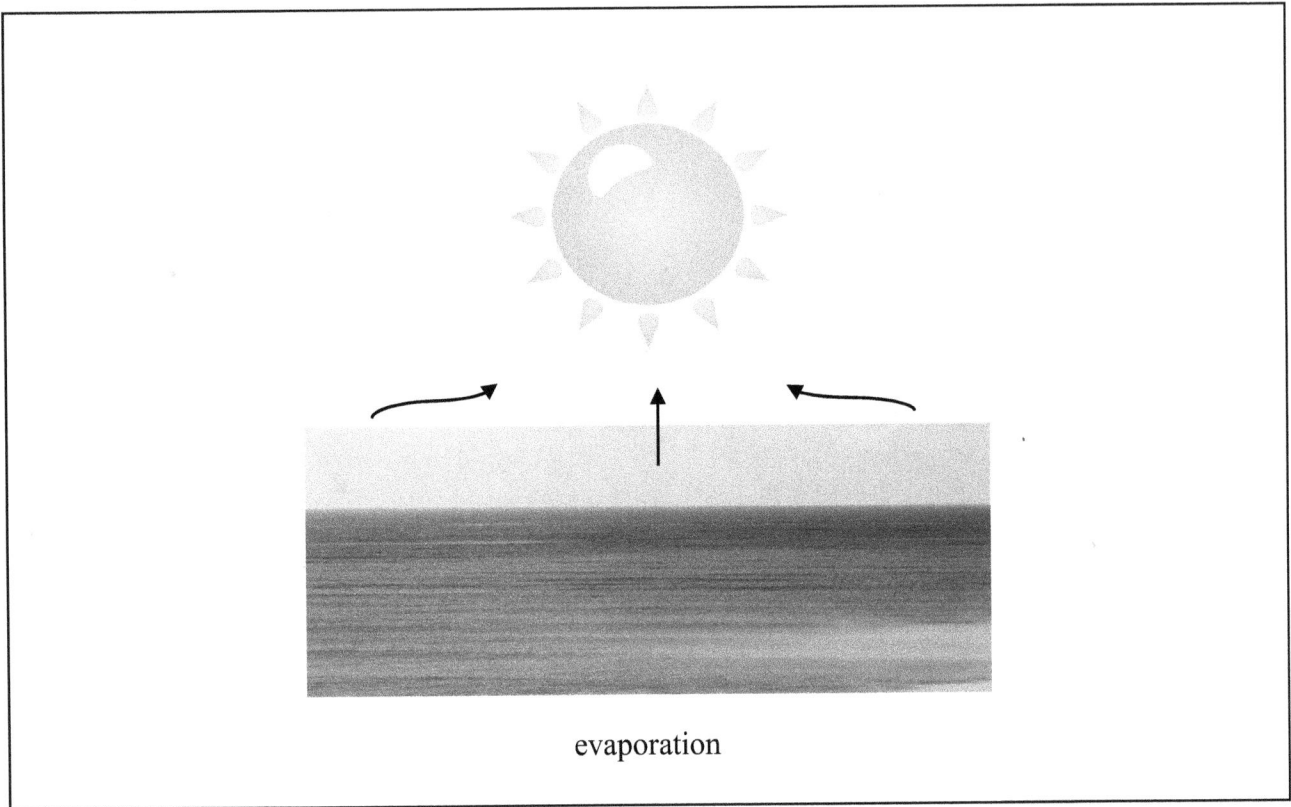

evaporation

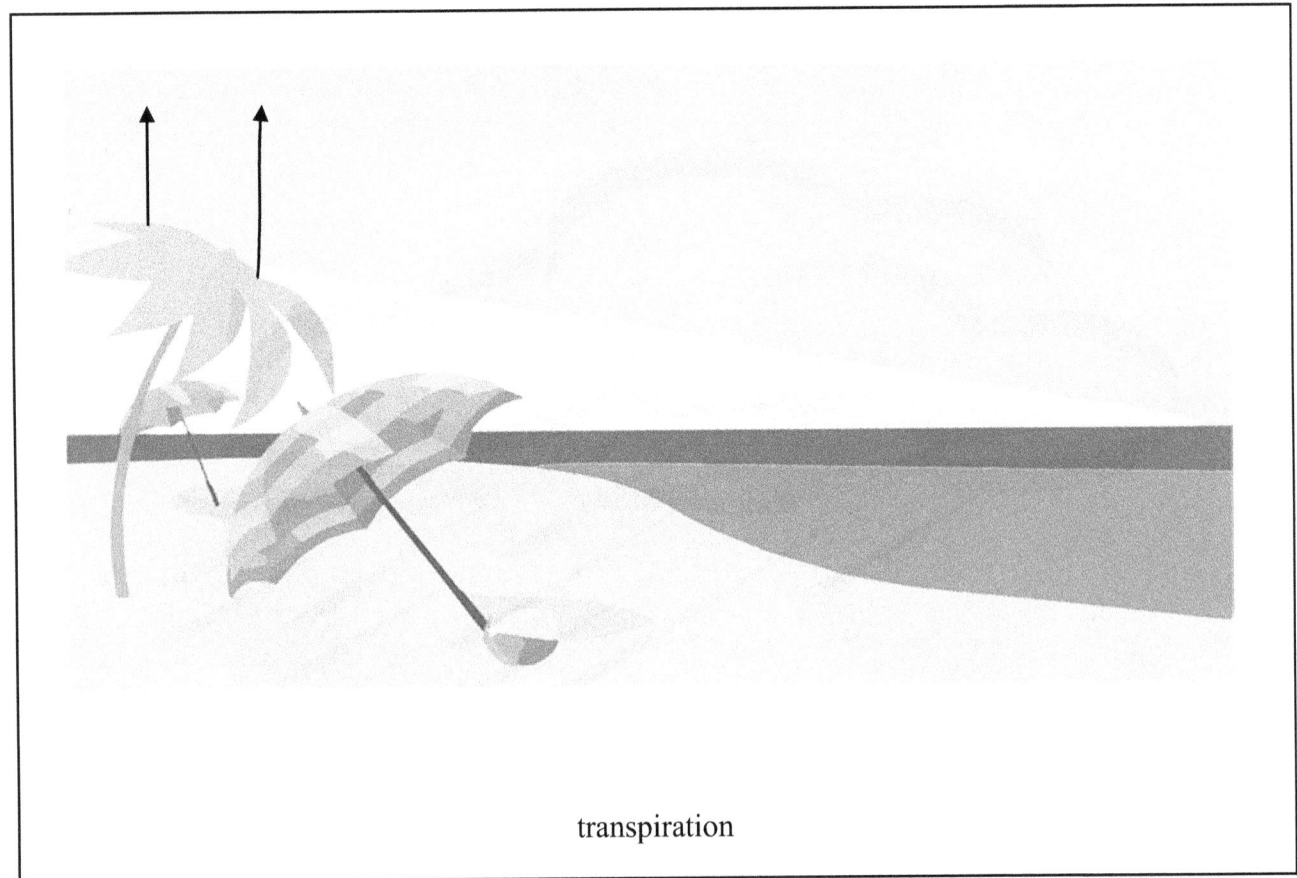

transpiration

run-off

The Big Water Puddle

Across the Curriculum Activities (Art/Science) Group Project

Group Project (3 days)

Materials:

computers   paper   markers

pencils   science boards   online pictures

Students will work in groups of 4 to research the steps of the water cycle and create a presentation on a science presentation board.

Day 1-Students will divide the research topics, and each will research their topic and write down the information (this will become a summary later for the project). Student 1 -evaporation, Student 2-condensation, Student 3-precipitation and Student 4-role of ocean and sun in the water cycle. Students will share their findings with the group.

Day 2-Students will work together to brainstorm- design how they intend to present the information on their science project. Students will either draw or print pictures to put on their presentation board. The board should include at minimal a tree, a body of water, sun, clouds. Students will label parts of the water cycle. The cycle should include arrows pointing into the direction of the flow of water. Next to each label, students will place their summaries

Day 3-Students will meet briefly to discuss how they will present their project. Students will make their presentations to the class.

The Big Water Puddle

Materials:

11 X 17 white paper   pencils

markers/coloring pencils

Students will design their own sequence map of the water cycle. Students will brainstorm the layout of their water cycle system. They will then illustrate/draw pictures of the water cycle (include arrows, labels for each part) and write a sentence describing what is happening in each step. Students will add color to their diagram and share it with others.

The Big Water Puddle

Across the Curriculum Activities (Art/Science) Diorama

Diorama

Materials:		
shoe box	cotton balls	glitter glue pens
cardboard	glue or tape	wide popsicle sticks
clear string	clear/blue beads	construction paper or paint
tissue paper	foam	

Students will create a 3D Model of the water cycle by creating a diorama. Students will either glue construction paper or paint inside a shoebox to create a background for their water cycle. They will glue cardboard inside the box to create pop out scenes and/or various layers. Students will choose various materials to represent the elements of the water cycle (clouds, sun, water, etc.) from materials such as using blue glitter glue to create water at the bottom of the shoebox and stretching out cotton balls to attach as clouds. Students will then do a write up/summary describing what is occurring in the diorama representing the water cycle. See examples below.

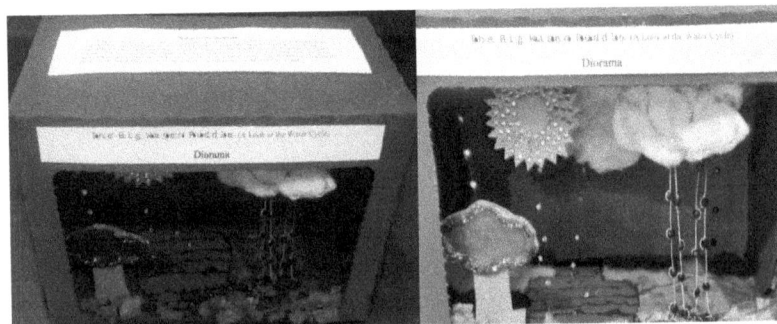

The Big Water Puddle

Across the Curriculum Activities (English Language Arts/Science)
Extension Activity & Writing Prompt

Extension Activity

Have students think about the story they just read or listened to, recall the characters. Suggest that students add a new character to the story. Tell what the name of the character would be and what part they would play in the story. What would be the significance of the additional character?

Writing Prompt

Have the students write to the prompt: Now that you have read the story, explain how you would teach the water cycle to someone else.

The Big Water Puddle

Resources

1. Florida Science Standards

<u>SC.5.E.7.1 & SC.5.E.7.2</u> can be found at the following website:

https://www.cpalms.org/Public/PreviewIdea/Preview/538

2. Florida ELA Standards

Strand: Writing Standard

Cluster 1: Text Types and Purposes LAFS.5.W.1.2 can be found at the following website:

https://www.cpalms.org/Public/PreviewStandard/Preview/5843

3. Strand: Language Standard

Cluster 3: Vocabulary Acquisition and Use LAFS.5.L.3.6 can be found at the following website: https://www.cpalms.org/Public/PreviewIdea/Preview/1678

1. Georgia Science Standard of Excellence

S4E3 can be found at the following website: https://www.georgiastandards.org/Georgia-Standards/Documents/Science-Fourth-Grade-Georgia-Standards.pdf

2. Georgia ELA Standards

Reading Literary, Key Ideas & Details-**ELAGSE4RL3** & Language, Vocabulary Acquisition and Use-**ELAGSE4L4** can be found at the following website:

https://www.georgiastandards.org/Georgia-Standards/Documents/Big-Book-Standards-ELA-and-Literacy-Standards.pdf

1. National Standards

Earth and Space Science

Content Standard D National Science Education Standards (1996) Ch 6 Science Content Standards can be found at the following website:

https://www.nap.edu/read/4962/chapter/8